I0510862

Plausible Scenarios for Thai Energy Businesses in the Next 30 Years

Nitida Nakapreecha
Jakapong Pongthanaisawan
Weerin Wangjiraniran

ELIVA PRESS

ELIVA PRESS

Nitida Nakapreecha

Jakapong Pongthanaisawan

Weerin Wangjiraniran

Energy sector is currently facing different challenges that are changing rapidly. Such factors include economic, social, environmental, technology and policies. Each of which affects the energy industry at different dimension and level. Some factors may deviate a business from achieving its goals, while some factors may impede business's future operations. Even worse, certain factors may have a serious impact on national energy security. Energy organization is, therefore, required to closely monitor and evaluate such changes, including exploring plausible scenarios in various contexts. The results will increase the organization's ability to tackle challenges and overcome obstacles in a firm and rational way.

This book presents plausible scenarios for Thailand's energy business under a sustainability perspective, but in different possible contexts. The findings will be useful in the analysis of national energy balance with detailed sector-by-sector projections. All of which will be beneficial in strategic energy planning at both the national and corporate level with a view to achieving the sustainable development goals (SDGs) by 2050.

Published by Eliva Press SRL
Address: MD-2060, bd.Cuza-Voda, 1/4, of. 21 Chişinău, Republica
Moldova
Email: info@elivapress.com
Website: www.elivapress.com

ISBN: 978-1-63648-135-7

© Eliva Press SRL, 2021
© Nitida Nakapreecha, Jakapong Pongthanaisawan, Weerin
Wangjiraniran
Cover Design: Eliva Press SRL
Cover Image: Freepik Premium

No part of this book may be reproduced or utilized in any form or by
any means, electronic or mechanical, including photocopying,
recording, or by any information storage and retrieval system,
without permission in writing from Eliva Press.

All rights reserved.

Plausible Scenarios for Thai Energy Businesses in the Next 30 Years*

N Nakapreecha[1,], J Pongthanaisawan[1], W Wangjirniran[1]**

[1] Energy Research Institute, Chulalongkorn University, Bangkok, Thailand

*This paper is an extended and revised article presented at the International Conference on Sustainable Energy and Green Technology 2019 (SEGT 2019) on 11-14 December 2019 in Bangkok, Thailand.

** **Correspondence:** nitida.n@chula.ac.th

Keywords: plausible scenarios, energy business, sustainable development goal, energy challenges, energy disruption, carbon policies, STEEP analysis.

Nakapreecha N, Pongthanaisawan J and Wangjiraniran W (2021) Plausible Scenarios for Thai Energy Businesses in the Next 30 Years. Front. Energy Res. 8:590932. doi: 10.3389/fenrg.2020.590932

Table of Contents

Abstract

Energy sector is currently facing different challenges. It is necessary for the business leaders to understand the forthcoming changes that will affect their businesses in order to prepare themselves for the uncertainties and take advantage of new opportunities. This paper identifies changes that are expected to have impacts on Thailand's energy system in the next 30 years and explores plausible scenarios of Thailand's energy business under such changes. The study starts with an examination of global and local circumstances. The examination pinpoints a focal issue as "to achieve the sustainable development goals (SDG) by 2050". Then, a tool so-called STEEP analysis is implemented to identify business drivers in social (S), technological (T), economical (E), ecological (E), and political (P) aspects. Subsequently, consultations with stakeholders are arranged to finalize the critical uncertainties. Policy and technology are found to be two of the most powerful factors affecting energy business and are, therefore, used as fundamental framework for scenario development. Accordingly, four plausible scenarios are derived providing different possible prospects of Thailand's energy business. The findings can further be used in the analysis of national energy balance with detailed sector-by-sector projections. All of which will be beneficial in the strategic energy planning at both national and corporate level with a view to achieving the SDGs by 2050.

1   Introduction

The energy sector is experiencing a series of changes. Such changes transform a stable and continual business environment into an uncertain and turbulent one. This poses great challenges to business investors in view of corporate strategic planning (Grant 2003). It is important for the business leaders to understand the forthcoming changes that will affect their businesses. Otherwise, they may have problems with adapting to those changes and will eventually have to pay considerable price in terms of value destruction, lower profits and decline in competitiveness. Several case studies have been observed.

In the 2000s, traditional utilities in Germany overlooked potential of distributed photovoltaic (DPV) generation. Later in the early 2010s, they lost 97 percent of their electricity generation market shares to the private investors (Richter 2013).

Moreover, recent studies of the International Energy Agency (IEA) show a fundamental shift away from fossil fuels to renewables in thermal power sector. The trend was observed from the fall of annual final investment decision (FID) for both coal and gas-fired plants during 2016-2018 (IEA 2019). The downturn in thermal power investment has seriously affected traditional thermal power producers and has led them to the painful solutions. General Electric (GE), Siemens and Toshiba are among the players dominating the sector. Nearly 74 percent of GE's market capitalization over 2016-2018 has collapsed from misreading the pace of the post-Paris energy transition (Buckley 2019), resulting in a cutting of about 12,000 jobs in power division (Proctor 2017). Likewise, Siemens had to spin off its power and gas division, while Toshiba had to reduce its coal-fired power business (IEA 2019).

On the other side, the changing context brings great business opportunities to those with agility to adapt to market and business environmental changes. As in the case of the declining thermal power investment, some utilities have merged their supply-side business to cover distribution networks and retails in order to ensure their supply resilience, energy efficiency, and professional demand-side services. For example, Enel, one of the energy giants, had acquired EnerNOC, an US-based leading provider of smart energy management services. This acquisition enables them to deliver the superior energy services to their current and prospective clients (EnerNOC 2017). Moreover, Enel had acquired eMotorWerks, a provider of e-mobility solutions, to be able to provide grid balancing solutions and pave their way into the promising electric mobility market. According to their head of Global eSolutions division, the acquisition fulfills their portfolio of grid flexibility services, which includes demand response network, distributed energy management systems and battery storage solutions (Enel 2016).

The above case studies emphasize the fact that an organization needs to keep an eye on the changing contexts – not just to survive, but to thrive in the business. Volatility of business environment makes strategic planning more difficult (Grant 2003). Thus, the business needs a systematic strategic foresight rather than intuition or gut feelings. They need a tool that can provide various possible future situations (scenarios). Such scenarios will expand the executives' visions beyond their experience and encourage creative thinking. As a result, the executive will be able to make more flexible strategies for their business.

In consideration of Thailand, the country needs energy to supply its domestic demand, which is in consistent with economic development, population growth and urban growth (Ministry of Energy 2015). Thus, Thailand has long given its priority to energy security. The statistical data shows energy consumption increases every year with high proportion of imported energy (Figure 1). The net import of commercial primary energy accounts for more than 50 percent of total supply since 1995. This proportion is anticipated to grow as the proven oil and gas reserves are depleting. This will affect not only energy security, but also the national energy expenditure (IRENA 2017).

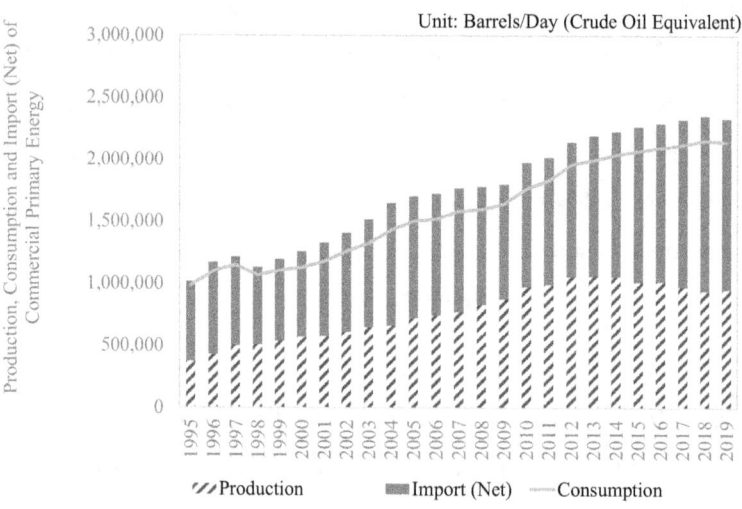

Figure 1. Production, consumption and import (net) of commercial primary energy

(EPPO 2020)

According to Figure 2 and 3, most of commercial primary energy consumption and import are fossil-based. This results in the increase of environmental problems, especially the increase of greenhouse gas (GHG) emissions. Additionally, people have negative attitude towards certain fossil fuels, especially coal. Protests against coal-fired power plants have been seen in the country (Heinrich-Böll-Stiftung 2020). In response, Thailand has set a renewable energy goal of 30 percent of total final energy

6

consumption by 2036 in the Alternative Energy Development Plan or AEDP 2015 with an intention to diversify energy sources as well as to decrease environmental pollutions (Ministry of Energy 2015).

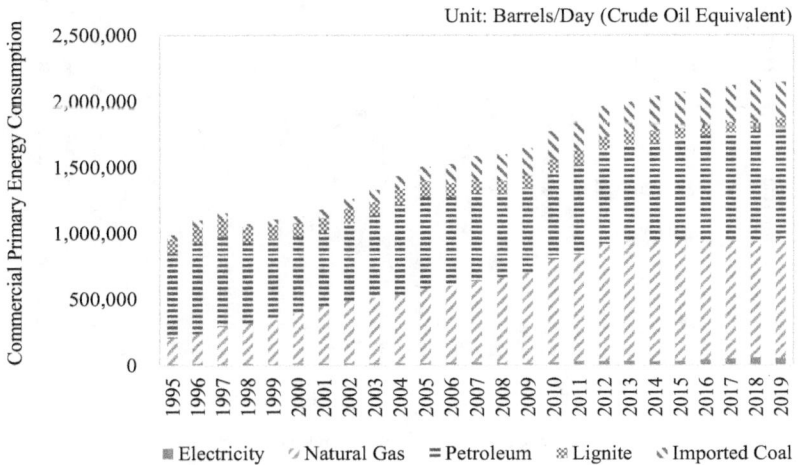

Figure 2. Commercial primary energy consumption

(EPPO 2020)

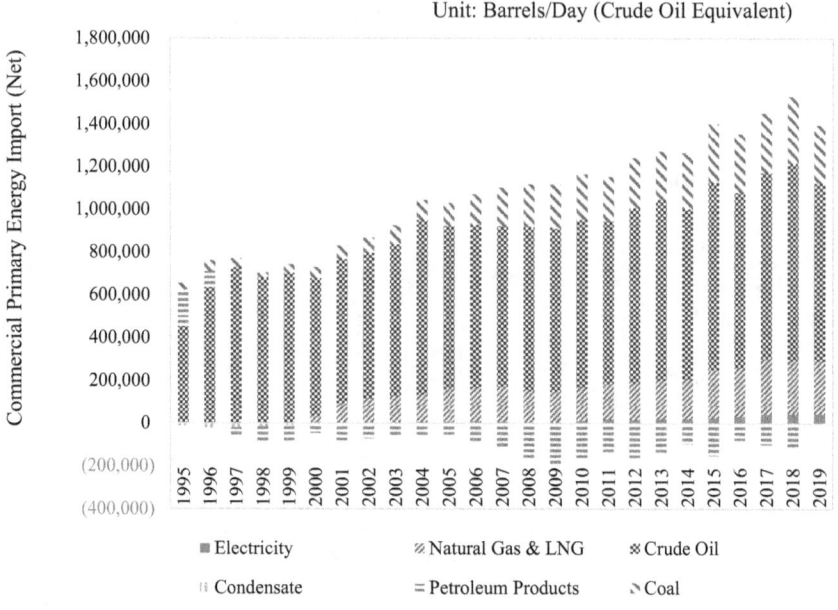

Figure 3. Commercial primary energy import (net)

(EPPO 2020)

In view of business players, Thai energy business is still centered around the state-owned enterprises (SOEs). In the power sector, the Electricity Generating Authority of Thailand (EGAT) is a SOE responsible for generating, transmitting and wholesaling electricity. In addition, there are two SOEs responsible for distributing and providing low-voltage electricity to end users. They are the Metropolitan Electricity Authority (MEA) and the Provincial Electricity Authority (PEA) (BOI 2020). Although Thailand has encouraged private companies to enter the power generating sector as an independent power producer (IPP), a small power producer (SPP) and a very small power producer (VSPP) in the power generation sector; electricity produced by these

8

private companies has to be sold to EGAT, PEA and MEA (BOI 2020). In the oil and gas sector, PTT public company limited, engages in the entire supply chain for oil and gas. Its businesses cover exploration, transmission, oil refinery and gas separation, and petrol station. It is the sole buyer of oil and gas produced domestically and the sole operator of gas transmission and separation (Nikomborirak 2017).

Considering the mentioned situations in Thailand, it can be anticipated that Thailand energy system will soon be facing the changing circumstances. Accordingly, it is important to Thailand to developed scenarios to foresee possible situations in order to be able to adapt to the upcoming changes.

Scenario analysis is a tool to envisage possible future events by examining various influential variables and identifying critical uncertainties. Scenarios are not a future prediction, but rather a visualization of alternative views of the future in the context of different sets of key variables (Grant 2003). This implies that scenarios are not limited to a single vision of the future, but they reveal a variety of possibilities. At the same time, they form a framework for describing pathway towards each vision and finding appropriate actions for each path (Schwartz 1996).

With the better understanding of future uncertainties, scenarios can reshape the framework people perceive the future (Gilovich 1981). And, as they reveal multiple possible future circumstances, scenarios have significant role in designing flexible strategies, which foster alertness and responsiveness of decision makers to those changing circumstances (Grant 2003). Unsurprisingly, several businesses have adopted the scenario planning in their strategic planning to explore and exploit the world of great uncertainties.

Shell is one of the world's well-known scenario developers. The company has been developing and deploying various scenarios since the 1970s to assist its executives in making decisions (Shell 2020). In 1970, its scenarios illustrated possibility of the oil price upsurge induced by the Organization of the Petroleum Exporting Countries (OPEC). Later in 1973, the oil prices actually quadrupled. Since then, Shell has been using scenario planning even more seriously to foresee and exploit succeeding oil price swing (Schoemaker 1993). In addition, the company has developed various scenarios to identify emerging global challenges, which have helped them make critical decisions in uncertain times and overcome energy and environmental challenges (Shell 2020).

The world energy outlook developed by the IEA is one of the most recognized energy outlooks that provides comprehensive scenarios, which map out the results of various energy policies and investment decisions. Their 2019 issue illustrates a pathway that enables the world to achieve climate, energy access and air quality targets while ensuring reliability and affordability of energy for an increasing population)IEA 2019).

Although, there are a number of global energy scenarios publicly available to be used as guidelines for exploring global energy outlook, the country still needs its own tailor-made scenarios specially built upon its own context. Moreover, it is important for the country to explore more than one possible scenario to avoid rigid strategic planning. This is to ensure that the formulation of energy strategies and policies will be appropriately developed for that country and will be flexible to changes.

In view of Thailand, there are 5 energy master plans, namely (1) Thailand Power Development Plan or PDP, (2) Energy Efficiency Development Plan or EEDP, (3) Alternative Energy Development Plan or AEDP, (4) Natural Gas Supply Plan, and (5) Petroleum Management Plan (EPPO 2016). These plans can provide framework for business strategy development. However, under the dynamic circumstances, it will be

10

too rigid for the business as well as the for the country's energy security to develop strategies based only on current plans without considering the upcoming changes.

This paper does not intend to indicate the future of energy business in Thailand. But it is to reveal changes that have impacts on Thailand's energy system in the next 30 years and explore plausible scenarios of Thailand's energy business under those changes. This is to provide narrative explanation of how Thailand's energy business can look like in the next 30 years. The result can be further useful to the assessment of future energy demand and supply as well as corresponding environmental impacts such as greenhouse emissions. All of which are essential to the formulation of effective energy strategies and policies at both corporate and national levels.

2   Methodology

The research consists of 4 steps: identification of focal issue, identification of drives, identification of critical uncertainties and development of plausible scenarios.

2.1 Identification of focal issue

All scenarios are developed based on the focal issue, which has critical consequences for the futures of the organization (Garvin and Levesque 2006). Thus, the first step is to identify the focal issue that an energy organization should focus on. For the effective business strategic planning, the focal issue should have a long-term influence on the business and be aligned with global changes. Therefore, it is important to address both Thailand's issues as well as global trends and development, which will affect future patterns of energy supply and consumption. Consequently, consultation with stakeholders from management level to practitioner level in a broad range of functions and divisions is arranged to finalize the focal issue. Timeframe, scope and focal variable are defined at this step.

2.2 Identification of drivers

At this step, key drivers that will significantly influence the focal issue are identified. A common tool for assessing influential factors so-called STEEP analysis is applied. The analysis provides an overview of current and future circumstances impacting the business in 5 areas: social, technological, economic, ecological and political (PESTLEAnalysis.com 2015). The framework of STEEP analysis is illustrated in Table 1. In the analysis, the issues and trends are mapped into the respective STEEP area to investigate how they affect the energy business.

Table 1 Framework of STEEP analysis (PESTLEAnalysis.com 2015)

Social	Technological	Economic	Ecological	Political
Social changes, e.g. demographic change, behavioral changes and lifestyle changes	Technological advancement and convergence, e.g. innovations and trends in product development	Changes in economic contexts, e.g. domestic economic growth, interest, income, profit and market competition	Ecological impacts of products and/or services in both physical and biological terms	Political changes affecting laws, regulations, and policies, e.g. national development plan, national energy plans*, incentives for businesses as well as regulatory

Social	Technological	Economic	Ecological	Political
				burdens for businesses

Remarks: *Include (1) Thailand Power Development Plan or PDP, (2) Energy Efficiency Development Plan or EEDP, (3) Alternative Energy Development Plan or AEDP, (4) Natural Gas Supply Plan, and (5) Petroleum Management Plan

It should be noted that not all the identified drivers are equally important or equally uncertain (European Foresight Platform 2010). In addition, some drivers are predetermined, while some are uncertain. Thus, it will be helpful to identify the inevitable and necessary ones as well as the unforeseeable and essential ones (Schwartz 1996). In this regard, further study on each driver is required to understand its essence and impact on the energy business.

2.3 Identification of critical uncertainties

In order to examine the influential drivers in detail, a workshop with stakeholders is organized to evaluated those drivers based on two criteria:1) the degree of importance for the success of the focal issue and 2) the degree of uncertainty underlying in those drivers (Schwartz 1996, Rialland A 2009). Consequently, the drivers are categorized into 4 groups (Figure 4).

- Group 1: The small significance (low importance, low uncertainty) is usually considered as insignificant to the scenario development and may be disregarded.

- Group 2: The revisit (low importance, high uncertainty) is the group that should be revisited periodically as its high uncertainty may alter its importance over time.
- Group 3: The importance (high importance, low uncertainty) should be addressed in all scenarios, owing to its high impact.
- Group 4: The critical uncertainty (high importance, high uncertainty) is the most critical driver in developing scenario logic and defining plausible scenarios.

The main purpose of this step is to identify the critical uncertainties as they are considered to have critical role in determining Thai energy scenarios. However, it should be noted that the real situation may be different from the developed scenarios. Nevertheless, it will encompass elements of the scenarios. Thus, the good scenario should identify the bound of the plausible futures. In addition to the key drivers, some other drivers with less importance and uncertainty should be integrated in the scenario building. They may be served as reference points, signal posts, narrative components, or minor details, which add more clarity to the scenarios (Rialland A 2009).

	Group 3 The Importance	Group 4 The Critical Uncertainty
Degree of Importance	Group 1 The Small Significance	Group 2 The Revisit

Degree of Uncertainty

Figure 4. Importance-uncertainty matrix(Rialland A 2009).

2.4 Development of plausible scenarios

The 2x2 matrix technique (Rhydderch 2017) is employed in developing scenarios in this study. In this regard, two of the critical uncertainties with greatest importance and uncertainty are selected to be used as fundamental axes of a scenario matrix. Determining these axes are one of the most important steps in the process of scenario development as it is necessary to ensure that the four generated scenarios are clearly differentiated and that their differences make a difference for decision making (Schwartz 1996, Rhydderch 2017). It is the step where intuition, insights and imagination play the biggest part (European Foresight Platform 2010). Subsequently, the scenario narratives are accordingly developed by applying other drivers previously identified.

3 Results and Discussion

3.1 Identification of focal issue

3.1.1 Global situation and challenge

According to the review, environment and digitalization are playing a leading role in the growth of global economic and energy businesses (World Economic Forum 2019). With the rise of environmental concerns, both of them have widely been integrated into economic and energy policies. The 3D's (decarbonization, decentralization and digitalization) have become even more prominent in shaping the directions of business, national and global policies (Silvestre, et al. 2018). Modern technologies and innovations are encouraged to be more and more environmentally friendly (Tundele 2015).

In addition, it is observed that the trend of energy demand will shift towards low-carbon options (Figure 5). Demand for renewable energy is expected to have the greatest growth rate making the renewables become one of the world's major energy sources over the next 10-20 years. In contrast, the future of fossil-based fuels is likely to be in the downtrend. Although it is forecasted that natural gas consumption will remain in the future, oil and coal consumptions are expected to decline dramatically, particularly in the advanced economy countries. Their growth will be seen only in the emerging markets.

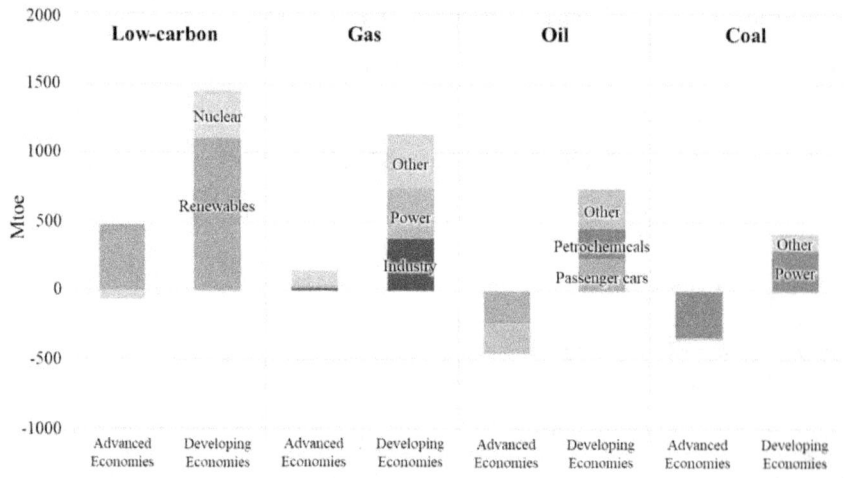

Figure 5. Change in total primary energy demand, 2017-40 in the new policies scenario (NPS)

(IEA 2018)

The above observation indicates the increase of environmental awareness globally. However, the most concrete movement showing that the world takes the environmental issues seriously is the establishment of sustainable development goals (SDGs).

Adopted by all United Nations Member States in 2015, the SDGs are recognized as a global call for action to eradicate poverty, protect the earth as well as ensure peace and prosperity for everyone by 2030 (UNDP 2020). Emphasizing the importance of these global goals, the IEA has proposed the sustainable development scenario (SDS) that maps out a way to meet sustainable energy goals across all parts of the energy system (IEA 2019). Moreover, Shell's Sky scenario illustrates a pathway to achieve the Paris Agreement target. It reveals the potential to evolve the energy system to meet global energy demand without delivering a climate burden to future generations (Shell International B V 2018).

3.1.2 Challenges for Thailand

Thailand is currently confronting several challenges impeding national development ranging from economic, social, and environmental (National Strategy Secretariat Office 2018). On the economic side, the economic structure has not yet been completely driven by innovative technology. While approximately 30 percent of total labor force are in the agricultural sector (National Statistical Office 2019), productivity of the sector is still relatively low. Furthermore, Thai workforces do not yet meet expected requirements and demands of the labor market. Unavoidably, poverty and income inequity remain the long-term concerns on the social side (National Strategy Secretariat Office 2018). In addition to economic restructuring, the country will need to stimulate new businesses that create higher values in order to improve national competitiveness, economic growth and income distribution. Energy security and stability are needed to support these needs.

Environmental issues have been the long-term problems in Thailand (Panya and Sirisai 2003). Air emissions are among the most concerned problems (Narita, et al. 2019). Carbon emissions from energy consumption are on the rising trend in major sectors, i.e. power generation, transport and industry (Figure 6). Thanks to national commitments to major international agreements, substantial action plans have been developed to tackle certain issues. Such commitments include the greenhouse gas reduction goals under the United Nations Framework Convention on Climate Change (UNFCC) and the sulfur limit for marine fuel oil set by International Maritime Organization (IMO) (ONEP 2015, MNRE 2015, Thai Oil 2020). However, local air pollution issues such as fine particulate matter ($PM_{2.5}$) are still waiting for serious and long-term solutions (Khidhir 2020).

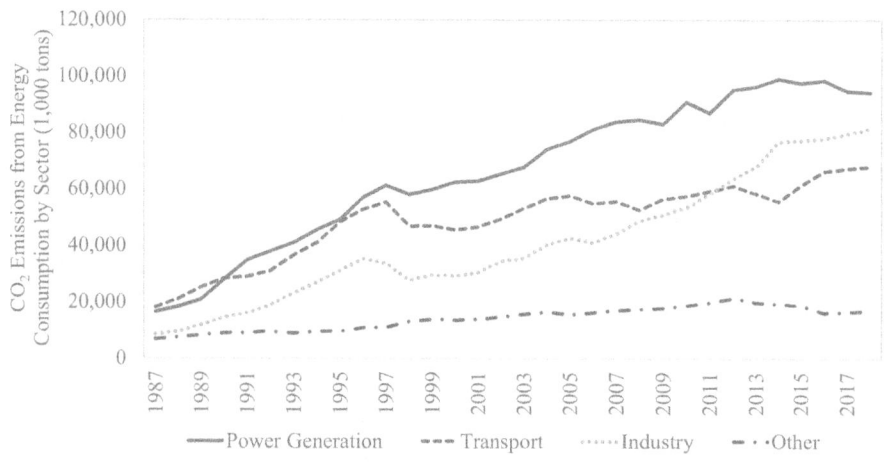

Figure 6. CO_2 emissions from energy consumption by sector

(EPPO 2020)

Thailand is expected to face the challenges of disruptive technologies (TDRI 2018). More importantly, the adoption of such technologies is anticipated to affect overall energy consumption pattern. Electric vehicle (EV) can be viewed as example. Previous study shows that EV users tend to recharge their vehicles at night (Pasaoglu, et al.

2013). As a result, night time load will increase. This is also in line with the solar PV case (Hudson and Heilscher 2012). Therefore, there is a chance that peak demand will be shifted from day to night with the growth of EV users and solar PV users. Preparations must be in place to cope with the demand shift, such as demand response programs, to avoid energy supply shortage.

The above findings suggest that, in the next phase of national development, Thailand needs to give priority to environmental protection in parallel with economic growth. As uninterrupted energy supply is a backbone of an economic development, the energy sector has to put a great effort to deliver reliable energy to those who need it without leaving environmental burden to the world. Accordingly, the focal issue that Thai energy business should focus on is "to achieve the sustainable development goals (SDG) by 2050".

3.2 Identification of drivers

Based on the STEEP analysis, factors which drive Thailand's energy system towards the SDG are described as follows.

3.2.1 Social factors

The shift in the lifestyle of Thai people towards a more urban and sharing economy is the first social factor identified for their influence on the future of energy business. Mobility-as-a-service is expected to become more popular as it offers more transport options, from busses to cars and bicycles (Sutthasri, et al. 2019). The advancement and convergence of digital technology will enhance efficiency of digital tools to facilitate the provision of this service. In addition, e-commerce is becoming even more popular as people will prefer electronic financial transactions through their personal gadgets (Rastogi 2018).

As people continue to migrate to metropolitan areas, a rise in population density is anticipated in certain areas, even though the overall population growth rate is decreasing (Vapattanawong and Prasartkul 2005). Aging society is another area that should be taken into consideration. Began in 2005, Thailand is expected to fully enter the aging society by 2021 with the population over 60 years of age accounting for more than 20 percent of the total population (Department of Elderly Affairs 2019). This results in a reduction in the overall number of labors and changes in consumer behavior. Both of which will have a negative impact on national economic development.

And lastly, a shift towards urbanization is anticipated. City development will be based on transit-oriented development (TOD) such as train and electric train. Such development will offer mobility advantages and flexibility as well as minimize reliance on personal vehicles (Nakamura, et al. 2016). Vertical living, particularly in the vicinity of public transport, will be increased (JICA; IDCJ; Pacet 2013).

3.2.2 Technological factors

The first technology factor expected to have an impact on Thailand's energy business is efficiency-enhancing technology, such as energy efficiency, product efficiency, production efficiency, and transport efficiency. It also covers zero energy designs and architecture (Ministry of Energy 2011).

Digital and information technology is most likely to increase its role in the energy business. With ability to provide prompt and accurate response, it is expected that this technology will be widely deployed in resource management system, such as smart grid, smart city and smart farming. All of which will lead to the greater energy and environmental conservation (IEA 2017, Virk, et al. 2020). Furthermore, the technology also helps improve people's quality of life. For example, high-speed internet and internet of things (IOT), which interconnect devices, provide an easy way for users to access, control and operate electric devices over the internet (Atkinson and Castro 2008, Farooq, et al. 2015). In addition, the digital and information technology will empower online market platforms as it enables safer and more efficient online transactions (Qin, et al. 2009).

The evolution of renewable and alternative energy technology has improved the production efficiency and has significantly driven the cost of renewable and alternative energy down (IRENA 2020). Consequently, this will improve price competitiveness of green energy against fossil fuels. And when combined with the higher efficient energy storage systems, the demand for renewable and alternative energy is expected to increase (IEA 2011).

Likewise, technology of electric and autonomous vehicles has been greatly developed so that major limitations are eradicated and vehicle prices are substantially reduced (Seba 2014). With a solid plan on the development of charging infrastructures (EGAT; PEA; MEA 2016), it is expected that the demand for the electric and autonomous vehicles in Thailand will be increased. Additionally, high-speed trains are also expected to have a larger role in the future (Ministry of Transport 2017). Robotics, artificial intelligence and advanced materials are also marked important in the development of energy system (Perez, et al. 2017, ACerS; AIST; ASM; MRS; TMS 2010).

3.2.3 Economic factors

Numerous economic factors are identified for their potential impacts on the energy business. National economic growth is one of the major influential factors as they obviously reflect national demand for energy and resources (Yeager, et al. 2012).

Certain Thailand's policies aiming to stimulate domestic income, trading power, and growth of industrial and service sectors (National Strategy Secretariat Office 2018) are also expected to affect total energy demand.

Different economic and industrial structures entail different energy intensity and energy consumption patterns. Heavy industry-based economy, which is generally highly resource intensive, will face the increase of energy consumption at an accelerated rate. On the contrary, economy that is driven by high value-added industries and services normally consumes comparatively less resources (ERI 2020). Thus, the country with this kind of economy and industrial structure will have greater overall energy efficiency.

Energy prices in global markets will continue being a key consideration for consumers in energy and technology selection as it is often seen that the markets quickly respond to price changes. In addition, the role of geopolitics as a main driver behind oil price changes is expected to be lessened, while unconventional energy are increasing its influence on the energy market (Neville, et al. 2017).

International economic cooperation provides options for international energy trading and energy security strengthening (Liu, Sheng and Azhgaliyeva 2019). Such cooperation includes the Association of Southeast Asian Nations (ASEAN), the framework of cooperation between ASEAN and the negotiating countries (ASEAN Plus), and the Economic Cooperation Development Program in the Greater Mekong Sub-regional (GMS).

Mega projects usually provoke energy demand in various forms as they entail the establishment of new industrial estates and the development of more connected transportation network.

Various sustainability concepts such as circular economy, bio-economy and sufficiency economy philosophy, are expected to have impacts on energy system. Described as an economy with closed-resource loop (Wautelet 2018), circular economy encourages rematerializing and reusing wastes with the aim to decouple economic growth from finite resource use (Ellen Macarthur Foundation 2017). In the other word, circular economy will be a key mechanism to promote waste utilization including waste-to-energy. This will affect the demand for virgin materials as well as production and consumption patterns (Rizos and Behrens 2017).

Bio-economy is expected to give a positive impact on the development of bio-energy. This is particularly for the use of agriculture waste and residual as energy. According to its purpose that is to promote self-sufficiency (The Chaipattana Foundation 2017), the philosophy of sufficiency economy can be applied in the transition of energy consumers to energy prosumers.

The last factor that is expected to have impact on energy business is economic crisis. Economic crisis is an event that is unlikely to occur but it will have an enormous impact on both economy and energy sector. The mid-1997 crisis had plunged Thailand's GDP growth to the lowest point since 1960 (Figure 7). The country's external debt had reached more than $70 billion or about 38 percent of GDP (Hewison 1999). And as a net energy importer, the increase of oil price, together with the increase of interest rates on debt from past infrastructure borrowing, has put a heavy burden on the country (Julian 2000). The economic stagnation had shrunk total energy demand and hampered new energy development projects as well as facility construction projects (Nakano and Honda 2000). The government had finally decided to privatize energy industry and make institutional arrangements with the objective to improve overall performance (Chaivongvilan, Sharm and and Sandu 2008). In addition, it is expected

23

that the economic crisis may have a negative impact on renewable projects as the country may need to turn to cheaper energies, such as coal for their cost minimization (Nakano and Honda 2000).

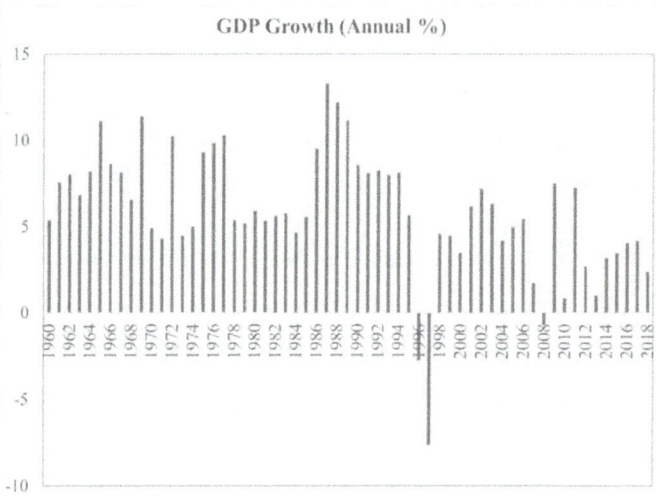

Figure 7. Thailand GDP growth

(World Bank 2019).

3.2.4 Ecological factors

Climate change agreement is an important mechanism for the continuous growth of renewable energy market (IRENA 2017). The commitment on greenhouse gas (GHG) emissions requires mitigation measures at global scale. Common measures include enhancing energy efficiency and fostering the use of renewable energy (IRENA 2017). Additionally, carbon pricing instruments, e.g. cap-and-trade and carbon tax, are also important drivers for GHG emission mitigation   As the carbon pricing may have adverse effects on carbon intensive industries (Kojima and Asakawa 2016), careful policy formulation is required to balance the effort to reduce GHG emissions and mitigate negative effects of the instruments.

Local air pollution issues such as fine particulate matter ($PM_{2.5}$) and nitrogen oxides (NOx) has become one of the most critical issues in Thailand during the past few years (Narita, et al. 2019). With the concern over their adverse health effects, public awareness of pollution reduction has been dramatically raised. The need to reduce such pollutions stimulates the consumption of low-carbon fuels and renewable energy in power, industry and transport sectors (Narita, et al. 2019).

As positive attitude of consumers towards green products and services are expected to be increased, producers and service providers need to pay more attention to the life cycle assessment. This also encourages organization and business enterprise to improve their corporate carbon footprint (Thongplew, Spaargaren and Koppen 2017).

Domestic reserves of natural resources, which include crude oil, natural gas, coal as well as renewable energy potential, are important considerations in the development of policies concerning energy usage in each sector. For example, the decline of domestic natural gas reserves has led to a shift in power sector fuel mix towards more fuel diversification with less natural gas proportion (Ministry of Energy 2015). In addition, based on the fact that Thai agricultural sector has high potential for renewable energy, the government has set the challenging target for biogas and biomass development (Suthiwong 2017).

3.2.5 Political factors

Several policies are considered to have potential impacts on the energy system of Thailand. The first one is legislations to promote renewable energy at national level. The economics of renewable energy are typically not competitive comparing to that of fossil fuels (Griffith-Jones, Ocampo and Spratt 2011). Moreover, renewable energy does not have prolonged government subsidies, unlike fossil fuels. These make renewable energy more expensive and more difficult to penetrate the market (ERI 2015). Thus, supporting legislations are needed to boost the market for renewable energy at the early stage when its price is still comparatively high. Example of such policies are feed-in tariffs and biofuel mandate (NESDB 2011).

Mandatory measures are effective tools to achieve desired outcomes. Current mandatory measures in Thai energy sector involve energy performance control, such as the requirement of energy conservation plan in designed buildings and designated factories, and the implementation of building energy code for building with total area equal to 2,000 square meters or more (Office of the Council of State 2007, Government Gazette 2009). In addition to Thai measures, some international regulations also have impact on energy businesses in Thailand. Example of the international regulations is the International Maritime Organization (IMO) regulation, which limit sulphur content in marine fuel oil at 0.5 percent mass by mass for vessels operating outside Emission Control Areas (ECAs) and at 0.1 percent mass by mass for vessels operating within ECAs (EIA 2019).

Policies concerning the development of energy business structure and governance have impacts on the development of energy efficiency, the determination of fuel type, and the growth of renewable energy (ERC 2018). Moving towards a competitive industrial structure with the use of advanced trading platforms will optimize fuel selection. In addition, realizing the benefits of liberalization in energy sector, Thailand has promoted competition in the oil business, while starting to allow the third-party access to pipeline infrastructure and LNG terminal in the natural gas business (ERC 2018).

To encourage self-dependency in the digital age, decentralization policies support the power development for self-consumption and peer-to-peer trading (IRENA 2020). It is expected that there will be local power development across the country as well as central infrastructure for inter region power management (Ministry of Energy 2018). In addition, according to the Power Development Plan 2018 (PDP2018), Thailand begins to assess regional power demand and supply as a part of the feasibility study on decentralization system development. Moreover, there are various pilot plants supporting decentralization, such as the smart grid pilot project in Mae Hong Son Province (EGAT 2018) and the solar roof power generation project (Energy Regulatory Commission 2016).

Energy pricing policies are one of the tools that government uses to set direction of energy development. They have been used to stimulate markets for renewable energy as well as other energy. Previous pricing policies that have been used include price fixing policies for natural gas Vehicle (NGV), adder and feed-in-tariff policies for renewable energy, and price subsidy policies for biofuels. Energy pricing policies are expected to become even more important, especially those that dynamically reflect actual cost and market prices such as real-time pricing policies (EPPO 2014).

3.3 Identification of critical uncertainties

Drivers identified in 3.2 are prioritized and assessed for their importance and uncertainties. The result is shown in Figure 8.

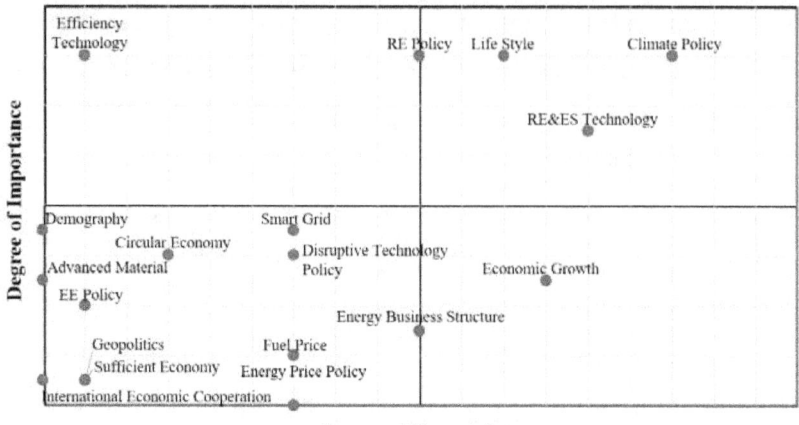

Figure 8. Ranking of drivers of Thai energy business towards
sustainable development goals

From Figure 8, drivers with high impact and high uncertainty or so-called "critical
uncertainties" are carbon policies, people lifestyle, renewable energy (RE) policies, and
renewable energy & energy storage (RE&ES) technologies. These four factors are
expected to have critical impacts on the direction of Thailand's energy business as they
reflect the trends of energy demand and supply as well as the urge to achieve
sustainable development goals. Accordingly, they should be taken into account in the
development of a national energy roadmap or master plan.

With high impact and low uncertainty, efficiency technologies should be included in
energy planning as well. The economic growth may be revisited periodically to observe
their impacts on the energy business. Other factors can also be parts of future
projection or may be disregarded.

3.4 Development of plausible scenarios

Focal issue, drivers and critical uncertainties identified in previous steps are used in developing plausible scenarios. Considering the critical uncertainties, factors that are crucial for Thai energy businesses to achieve the sustainable development goal within 2050 can be categorized into 2 groups. The first one relates to the advancement of technologies especially the disruptive technology, renewable energy technology and energy storage system. The other one involves policy-driven factors, especially renewable energy policies and climate policies. Both groups will be the significant driving forces in changing customer behavior and promoting the adoption of eco-friendly technologies leading to the achievement of sustainable development goals.

Accordingly, four plausible scenarios of "to achieve the sustainable development goal (SDG) by 2050" are developed as shown in Figure 9.

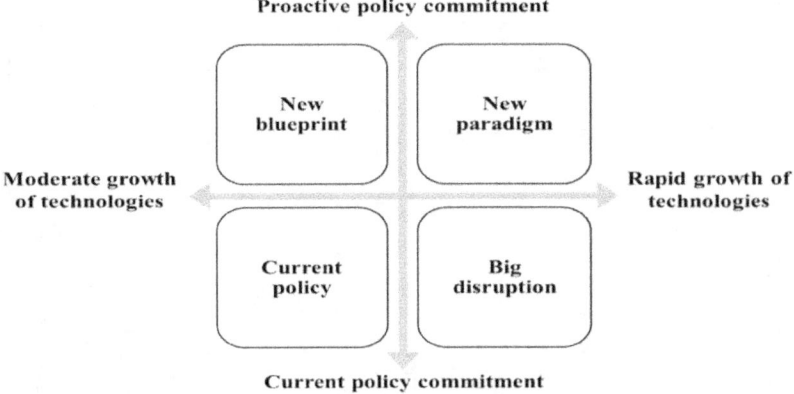

Figure 9. Plausible scenarios of Thai energy business in the next 30 years

Narratives of the four scenarios on the perspective of social, technological, economic, ecological and political are shown in Table 2

Table 2. Narratives of plausible scenarios for Thai energy business towards the SDG in 2050

Perspective	Current policy	Big disruption	New blueprint	New paradigm
Social	• People, especially those in urban, will gradually change their lifestyle towards digital lifestyle. • Environmental awareness will be growing due to severe pollution problems.	• People will shift their lifestyles towards digital lifestyle with greater environmental awareness and concern, influenced by the availability of clean technologies.	• People will shift their lifestyles towards digital lifestyle with greater environmental awareness and concern, influenced by policy incentives and regulations.	• People will shift their lifestyles towards digital lifestyle with serious concern on environment. • Eco-system for clean options will be ready for people to make choices.
Technology	• Clean technology will be adopted by limited group of people and corporate social enterprise. • Only distributed photovoltaics (DPV) and electric vehicles	• Clean technology will be available at competitive cost. Nevertheless, the penetration of new technology will be limited due to regulatory constraints.	• Growth of clean technology will be driven mainly by government policy. Direct and indirect incentives will be implemented to	• Clean and disruptive technologies will be widely adopted in diverse applications throughout daily life. • Smart city concept and energy disruptive technologies will be

Perspective	Current policy	Big disruption	New blueprint	New paradigm
	(EV) will thrive commercially in the domestic market due to their competitive costs.	• The DPV, EV and other potential technologies will be able to penetrate the domestic market.	promote clean energy technologies. • Smart cities with clean energy technologies will be seen in demonstration projects invested by local and central governments.	widely adopted in major cities across the country. • Integrations of smart energy solutions will be found in the development of business and industry. Examples of such integrations are the integration of DPV with energy storage system (ESS) and peer-to-peer (P2P) network, and the integration of EV or connected & automated vehicles (CAV) with vehicle-to-everything (V2X).

Perspective	Current policy	Big disruption	New blueprint	New paradigm
Economic	• National economy will be driven by tourism and export-led industrial production. • Petroleum and petrochemical companies will still hold outstanding market capitalization. • State enterprises will play a major role in the energy and supply management sectors.	• Several businesses, particularly petroleum and auto-part industries, will face the risk of digital disruption. • Due to the high penetration of DPV & EV, state enterprises will face challenges on load management and energy pricing.	• Digitalization will become a major added value for the national economy • Energy business will be gradually transformed towards decentralization. • The number of business transformation will be limited due to a lack of technology support.	• Digitalization and clean disruption will contribute significantly to the national economy. • Smart and low carbon cities will become new standard for major cities. • Energy business will be completely transformed towards decentralization with available prosumer business model. • Market platform will be initiated with the objective to create competition which will

Perspective	Current policy	Big disruption	New blueprint	New paradigm
				lead to efficiency maximization and cost minimization.
Environment	• Pollutions from economic activities will still be growing in the long run but at a slower pace due to the efficiency and management improvement.	• Penetration of clean technologies will lessen environmental problems but at a limited capability.	• Decoupling of economic growth from pollution generation will be seen before 2050.	• GHG emission reduction for the well-below-2-degrees (2DC) target will be achieved. • Both public and business sectors will engage in pollution and waste reduction.
Politics	• Climate change and environmental regulations will become part of national energy policy without any structural	• Similar to the current policy scenario, climate change and environmental regulations will become part of national energy policy	• The SDG and 2DC targets will become part of national agenda by 2050. • Constraints for prosumer and peer-to-	• The SDG and 2DC targets will become part of national agenda by 2050 with strong commitments from key stakeholders

33

Perspective	Current policy	Big disruption	New blueprint	New paradigm
	reform implementation.	without any structural reform implementation.	peer energy trading will be unlocked.	• Various economic tools will be implemented, e.g., carbon tax and road tax.

34

The above four scenarios indicate different situations, which may occur in the future. They are differentiated by the degree of policy and technology development. The current policy scenario best reflects the most possible view of Thai energy business. The scenario is developed on current situations with consideration of mega-trends as well as changes in economy, society, environment and technology. The goals of national energy plans are also taken into account based on current structure of energy business. It is foreseen that new technologies are unable to hold much of their market share under this scenario.

The big disruption scenario indicates quick market responses to new technologies, while there are no obvious changes or preparedness in business policies and structures. On the other hand, the new blueprint scenario refers to the situation, in which significant improvements have been made to energy business policies and structures but there is not much development in emerging technologies.

The new paradigm scenario describes that advanced energy technologies are broadly used and the country has favorable policies to support the development of energy industry. It is, therefore, the most desirable prospect as it best reflects the achievement of sustainable development goals. In this scenario, the energy sector is equipped with innovative technologies including disruptive technologies, advanced renewable energy technologies and energy storage systems. Additionally, there is policy support stemming from the momentum of climate change agreements and policy structural changes to support the growth of clean energy technologies.

In view of new business potential, different scenario entails different new business potential depend on policy and technology availability. The new paradigm scenario is expected to offer high market potential with available technologies and policy supports. However, without policy improvement or technology advancement, the business opportunities will be limited to existing technologies.

The developed scenarios can be further used in the development of an energy model, from which overall national energy balance can be analyzed with detailed sector-by-sector projections. The analysis will provide correlation of energy consumption, energy supply and environmental loads. The results will be useful in planning the procurement of clean energy in an appropriate proportion to Thailand 's energy system with a view to achieving the SDGs by 2050.

4   Conclusion

Various trends and challenges, both globally and locally, indicate the importance of sustainable development that needs an integrated approach to promote economic growth without compromising environmental protection. Likewise, the energy business should incorporate sustainable development into their corporate vision and strategy in order to successfully thrive in the face of new challenges. In this regard, the study proposes "to achieve the sustainable development goals (SDGs) in 2050" as a focal issue for Thailand's energy business.

Among diverse factors, policy and technology are found to be two of the most powerful factors to achieve the goal. Four plausible scenarios derived from these two factors explain alternative possible situations of Thailand's energy business in the next 30 years. Each of which shows different context of energy business according to the level of policy and technology development.

The current policy scenario is the most likely prospect under Thailand's current circumstances. It illustrates that new technologies are still unable to hold much of their market share under existing policies and plans. However, there are also other possible scenarios that deviate from the current policy scenario. The new blueprint scenario reflects the government's proactiveness in developing policies to support the SDGs. On the other hand, the big disruption scenario reflects the delay of policy supports, thus is considered to be the least desirable prospect as the national energy management system may not be ready for the disruptive technologies. The new paradigm scenario is the most desirable one as there will be both policy and technology available to support the SDGs achievement. Undoubtedly, business models under the new paradigm scenario will be completely different from the current ones.

Unanswered questions still exist for the energy sector, including how to effectively reallocate capital investments in order to achieve the SDGs under the ongoing energy transition. Therefore, further study on energy modelling is recommended to assess primary and final energy demand as well as the corresponding environmental loads. The results will be useful in the strategic energy planning at both national and corporate level with a vision to achieve the sustainable development goals by 2050.

5   References

ACerS; AIST; ASM; MRS; TMS. 2010. "Advanced Materials for Our Energy Future." Compiled by Materials Research Society.

Atkinson, R.D., and D.D. Castro. 2008. *Digital Quality of Life: Understand the Personal & Social Benefits of the Information Technology Revolution.* Washington.

BOI. 2020. *Electricity.* July 31. Accessed October 2020. https://www.boi.go.th/index.php?page=electricity.

Buckley, T., Hipple, K., Sanzillo, T. 2019. "General Electric Misread the Energy Transition: A Cautionary Tale." *Institute for Energy Economics and Financial*

Analysis. June. https://ieefa.org/wp-content/uploads/2019/06/General-Electric-Misread-the-Energy-Transition_June-2019.pdf.

Chaivongvilan, S., D. Sharm, and S. and Sandu. 2008. "Energy Challenges for Thailand: An Overview." *GMSARN International Journal* 2: 53-60.

Department of Elderly Affairs. 2019. "Measures to Drive the National Agenda on Aging Society (Revised Version)." Bangkok: Amarin Printing and Publishing. 9.

EGAT. 2018. *EGAT – PEA signs MOU for developing smart grid project in Mae Hong Son Province.* November 6. https://www.egat.co.th/en/news-announcement/news-release/egat-pea-signs-mou-for-developing-smart-grid-project-in-mae-hong-son-province.

EGAT; PEA; MEA. 2016. "Infrastructure Development Plan to Support Electric Vehicles in Thailand." Electricity Generating Authority Thailand; Provincial Electricity Authority; Metropolitan Electricity Authority.

EIA. 2016. "Chapter 7: Industrial sector energy consumption." In *International Energy Outlook 2016 with Projections to 2040*. Washington, D.C.: U.S. Energy Information Administration.

—. 2019. "The Effects of Changes to Marine Fuel Sulfur Limits in 2020 on Energy Markets." *Independent Statistics & Analysis.* Washington, DC: Energy Information Administration, U.S. Department of Energy, March.

Ellen Macarthur Foundation. 2017. *The Circular Economy in Detail .* Accessed October 2020. https://www.ellenmacarthurfoundation.org/explore/the-circular-economy-in-detail.

Enel. 2016. "Enel Press Release." *Enel.* April. https://www.enel.com/content/dam/enel-common/press/en/1671686-1_PDF-1.pdf.

Energy Regulatory Commission. 2016. "Announcement of the Energy Regulatory Commission regarding the Pilot Project on the Production of Solar Power on the Roof Freely."

EnerNOC. 2017. "Enel Group Completes Acquisition of Leading Us-Based Provider of Smart Energy." *Enel X.* August 7. https://www.enelx.com/content/dam/enel-x-na/press-releases/2017/ENOC_News_2017_8_7_General_Releases_0.pdf.

EPPO. 2014. *Annual Report 2014.* Annual Report, Bangkok: Energy Policy and Planning Office.

—. 2020. *CO2 Statistic.* http://www.eppo.go.th/index.php/en/en-energystatistics/co2-statistic.

—. 2020. *Summary Statistics.* Accessed October 2020. http://www.eppo.go.th/index.php/en/en-energystatistics/summary-statistic?orders[publishUp]=publishUp&issearch=1.

—. 2016. "Thailand Integrated Energy Blueprint." *EPPO Journal Special Issue 2016.* Bangkok: Energy Policy and Planning Office.

ERC. 2018. "Energy Regulatory Strategy No.3 (2018-2021)." Bangkok: Energy Regulatory Commission.

ERI. 2015. *Scaling Up Solar PV: A Roadmap for Thailand.* Department of Alternative Energy Development and Efficiency; Chulalongkorn University; British Embassy Bangkok, Energy Research Institute.

ERI. 2020. *Scenario Analysis and Modeling for Business Strategies.* Bangkok: Energy
    Research Institute.

European Foresight Platform. 2010. *Scenario Method.* Accessed March 2020.
    http://www.foresight-platform.eu/community/forlearn/how-to-do-
    foresight/methods/scenario/.

Farooq, M., M. Waseem, S. Mazhar, and A. Khairi. 2015. "A Review on Internet of
    Things (IoT)." *International Journal of Computer Applications* 113: 1-7.
    doi:10.5120/19787-1571.

Garvin, D.A., and L.C. Levesque. 2006. "A Note on Scenario Planning." Boston: Harvard
    Business School Publishing, July 31.

Gilovich, T. 1981. "Seeing the Past in the Present: The Effect of Associations to Familiar
    Events on Judgements and Decisions." *Journal of Personality and Social
    Psychology* 797-808. doi:10.1037/0022-3514.40.5.797.

Government Gazette. 2009. "Ministerial Regulation on Ministerial Regulation
    Prescribing Type or Size of Buildings and their Standards, Criteria and Methods
    for Building Design for Energy Conservation, B.E. 2552 B.E." Vol. 126. Bangkok,
    February 20.

—. 1995. "Royal Decree on Designated Buildings ." Vol. 112. Bangkok, August 14.

—. 1997. "Royal Decree on Designated Factory B.E.2540 (1997)." Vol. 114. Bangkoko,
    March 19.

Grant, R.M. 2003. "Strategic planning in a turbulent environment: Evidence from the oil majors." *Strategic Management Journal* (John Wiley & Sons, Ltd.) 24: 491-517. doi:10.1002/smj.314.

Griffith-Jones, S., J.A. Ocampo, and S. Spratt. 2011. "Financing Renewable Energy in Developing Countries: Mechanisms and Responsibilities." European Report on Development.

Heinrich-Böll-Stiftung. 2020. *The Coal Situation in Thailand and Strategic Environmental Assessment.* Accessed October 2020. https://th.boell.org/en/2018/05/11/coal-situation-thailand-and-strategic-environmental-assessment.

Hewison, K. 1999. "Thailand's Capitalism: The Impact of the Economic Crisis." *UNEAC Asia Papers.*

Hudson, R., and G. Heilscher. 2012. "PV Grid Integration – System Management Issues and Utility Concerns." *Energy Procedia* (Elsevier) 25: 82 – 92. doi:10.1016/j.egypro.2012.07.012.

IEA. 2019. *Coal-fired power generation capacity subject to a FID by region (annual average), 2010-2017.* December 9. Accessed July 2020. https://www.iea.org/data-and-statistics/charts/coal-fired-power-generation-capacity-subject-to-a-fid-by-region-annual-average-2010-2017.

—. 2017. "Digitalization & Energy." International Energy Agency.

—. 2019. *First steps in a reshaping of the energy industry landscape.* Accessed June 2020. https://www.iea.org/commentaries/first-steps-in-a-reshaping-of-the-energy-industry-landscape.

—. 2011. *OECD Green Growth Studies: Energy.* International Energy Agency.

IEA. 2019. *World Energy Outlook.* Paris: International Energy Agency. https://www.iea.org/reports/world-energy-outlook-2019.

IEA. 2018. *World Energy Outlook 2018.* International Energy Agency. Accessed 2019.

—. 2019. *World Energy Outlook 2019.* November. https://www.iea.org/reports/world-energy-outlook-2019.

IRENA. 2017. "Climate Policy Drives Shift to Renewable Energy." *International Renewable Energy Agency.* Accessed October 2020. https://www.irena.org/-/media/Files/IRENA/Agency/Topics/Climate-Change/IRENA_Climate_policy_2017.pdf.

—. 2020. *Global Renewables Outlook: Energy Transformation 2050.* Abu Dhabi: International Renewable Energy Agency.

—. 2020. "Innovation Landscape Brief: Peer-to-peer Electricity Trading." Abu Dhabi.: International Renewable Energy Agency.

—. 2017. *Renewable Energy Outlook Thailand.* Abu Dhabi: International Renewable Energy Agency.

JICA; IDCJ; Pacet . 2013. "Data Collection Survey on Housing Sector in Thailand." National Housing Authority.

Julian, C.C. 2000. "The impact of the Asian economic crisis in Thailand." *Managerial Finance* 26: 39-48.

Khidhir, S. 2020. *Suffocating in Thailand.* January 30. Accessed October 2020. https://theaseanpost.com/article/suffocating-thailand.

Kojima, S., and K. Asakawa. 2016. "Carbon pricing:: a key instrument to facilitate low carbon transition." Institute for Global Environmental Strategies, July.

Liu, Y., Z. Sheng, and D. Azhgaliyeva. 2019. "Toward Energy Security in ASEAN: Impacts of Regional Integration, Renewables, and Energy Efficiency." Asian Development Bank Institute.

Ministry of Energy. 2015. "Alternative Energy Development Plan (AEDP2015)." Bangkok: Department of Renewable Energy Development and Energy Efficiency, September.

—. 2015. "Gas Plan 2015." *Energy Policy and Planning Office.* October. http://www.eppo.go.th/images/POLICY/PDF/Gas%20Plan%20_Final_Publish.pdf.

—. 2011. "Thailand 20-Year Energy Efficiency Development Plan (2011-2030)."

—. 2018. "Thailand Power Development Plan 2018-2037 (PDP2018)." Bangkok.

Ministry of Transport. 2017. "4-Year Performance of Ministry of Transport for Happiness of Thai People." Bangkok.

MNRE. 2015. "Climate Change Master Plan (2015-2050)." Ministry of Natural Resources and Environment, July.

Nakamura, K,, F. Gu, V. Wasuntarasook, V. Vichiensan, and H. Yoshitsugu. 2016. "Failure of Transit-Oriented Development in Bangkok from a Quality of Life Perspective." *Asian Transport Studies* 194-209. doi:10.11175/eastsats.4.194.

Nakano, K., and K. Honda. 2000. "Impacts of Financial Crisis on Asian Energy Supply and Demand and Outlook." *Energy in Japan* (161). Accessed October 2020. https://eneken.ieej.or.jp/data/en/data/old/pdf/e16102.pdf.

Narita, D., N.T.K. Oanh, K. Sato, M. Huo, D.A. Permadi, N.N.H. Chi, T. Ratanajaratroj, and I Prwamart. 2019. "Pollution Characteristics and Policy Actions on Fine Particulate Matter in a Growing Asian Economy: The Case of Bangkok Metropolitan Region." *Atmosphere* 10 (227). doi:10.3390/atmos10050227.

National Statistical Office. 2019. "Summary of the survey: working conditions of Thai workforce." *National Statistical Office.* Accessed July 2020. http://www.nso.go.th/sites/2014/DocLib13/%E0%B8%94%E0%B9%89%E0%B8% B2%E0%B8%99%E0%B8%AA%E0%B8%B1%E0%B8%87%E0%B8%84%E0%B8%A1/ %E0%B8%AA%E0%B8%B2%E0%B8%82%E0%B8%B2%E0%B9%81%E0%B8%A3%E 0%B8%87%E0%B8%87%E0%B8%B2%E0%B8%99/%E0%B8%A0%E0%B8%B2%E0% B8%A7%E0%B8%B0%E0%B.

National Strategy Secretariat Office. 2018. "National Strategy 2018 - 2037 (Summary)." Bangkok: Office of National Economic and Social Development Board.

NESDB. 2011. *Thailand: Clean Energy for Green Low-carbon Growth.* National Economic and Social Development Board.

Neville, K., J. Baka, S. Gamper-Rabindran, K. Bakker, S. Andreasson, A. Vengosh, A. Lin, J. Nem Singh, and E. Weinthal. 2017. "Debating Unconventional Energy: Social, Political, and Economic Implications." *Annual Review of Environment and Resources* 42. doi:10.1146/annurev-environ-102016-061102.

Nikomborirak, D. 2017. *Chapter 6: SOE Reform in Thailand: Preparing for Free Trade Agreements.* Thailand Development Research Institute, Bangkok: JETRO Bangkok/IDE-JETRO. Accessed October 2020.

Office of the Council of State. 2007. "Energy Conservation Act B.E.2535 (amended)."

ONEP. 2015. "Thailand's Intended Nationally Determined Contribution (INDC)." Bangkok: Office of Natural Resources and Environmental Policy and Planning, October 1. doi:https://www4.unfccc.int/sites/ndcstaging/PublishedDocuments/Thailand%20First/Thailand_INDC.pdf.

Panya, O, and S Sirisai. 2003. "Environmental Conciousness in Thailand: Contesting Maps of Eco-Conscious Minds." *Southeast Asian Studies* 41 (1). https://kyoto-seas.org/pdf/41/1/410105.pdf.

Pasaoglu, G., D. Fiorello, L. Zani, A. Martino, A. Zubaryeva, and C. Thiel. 2013. *Projections for Electric Vehicle Load Profiles in Europe Based on Travel Survey Data.* JRC Scientific and Policy Reports, European Commission; Joint Research Centre; Institute for Institute for Energy and Transport, Milan: Joint Research Centre of the European Commission. doi:10.2790/24108.

Perez, J.A., F. Deligianni, D. Ravi, and G. Yang. 2017. "Artificial Intelligence and Robotics." June.

PESTLEAnalysis.com. 2015. *What is STEEP Analysis?* https://pestleanalysis.com/what-is-steep-analysis/.

Proctor, D. 2017. *GE Cutting 12,000 Jobs in Power Division.* December 7. https://www.powermag.com/ge-cutting-12000-jobs-in-power-division/.

Qin, Z., L. Shundong, H. Yi, D. Jinchun, Y. Lixiang, and Q. Jun. 2009. "Security
    Technologies in E-commerce." In *Introduction to E-commerce*, by Z. Qin.
    Springer, Berlin, Heidelberg. doi:https://doi.org/10.1007/978-3-540-49645-8_4.

Rastogi, V. 2018. "Thailand's E-Commerce Landscape: Trends and Opportunities."
    Dezan Shira & Associates, July 6.

Rhydderch, A. 2017. "Scenario Building: The 2x2 Matrix Technique." Edited by K.
    Radford. *Futuribles International.*

Rialland A, and Wold K E . IGLO-MP2020 Working Paper No. 10-2009. 2009. "Future
    Studies, Foresight and Scenarios as Basis for Better Strategic Decisions."
    Innovation in Global Maritime Production 2020 Project.
    http://www.forschungsnetzwerk.at/downloadpub/IGLO_WP2009-
    10_Scenarios.pdf.

Richter, M 2013 . 55. 2013. "German Utilities and Distributed PV: How to Overcome
    Barriers to Business Model Innovation Renewable Energy." *Renewable Energy*
    (Elsevier) 55: 456-466.

Rizos, V., Tuokko, K., and A. Behrens. 2017. "The Circular Economy: A Review of
    Definition, Process and Impacts." CEPS Research Report No. 2017/8, Centre for
    European Policy Studies, Brussels. Accessed October 2020.
    http://aei.pitt.edu/85892/1/RR2017-08_CircularEconomy_0.pdf.

Schoemaker, P.J.H. 1993. "Multiple scenario development: Its conceptual and
    behavioral foundatiojn." *Strategic Management Journal* (John Wiley & Sons, Ltd.)
    14: 193-213.

Schwartz, P. 1996. *The Art of teh Long View: Planning for the Future in an Uncertain World.* New York: Doubleday.

Seba, T. 2014. *Clean Disruption of Energy and Transportation: How Silicon Valley Make Oil, Nuclear, Natural Gas, and Coal Obsolete by 2030.* California.

Shell International B V. 2018. *Sky Scenarios: Meeting the Goals of the Paris Agreement.* Accessed 2019. www.shell.com/skyscenario.

Shell. 2020. *Shell Scenarios.* https://www.shell.com/energy-and-innovation/the-energy-future/scenarios.html.

Silvestre, M.L.D., S. Favuzza, E.R. Sanseverino, and G. Zizzo. 2018. "How Decarbonization, Digitalization and Decentralization are Changing Key Power Infrastructures." *Renewable and Sustainable Energy Reviews* (Elsevier).

Suthiwong, R. 2017. *Renewable Energy for National Security.* Bangkok: National Defence College of Thailand.

Sutthasri, P, N. Phongluengthum, W. Malsukum, and N. Charoonpipatkul. 2019. *Sharing Economy: Implications for Thai Economy.* Bangkok: Bank of Thailand.

TDRI. 2018. *Reorienting the Thai Economy to Prepare for the Age of Technological Disruptions.* May 18. Accessed October 2020. https://tdri.or.th/en/2018/05/annual-con-2018-news/.

Thai Oil. 2020. "Presentation to Investors." *Thai Oil Public Company Limited.* March. Accessed October 2020. https://investor.thaioilgroup.com/misc/PRESN/20200330-top-MonthlyPresentation-202003.pdf.

The Chaipattana Foundation. 2017. *Philosophy of Sufficiency Economy.* Accessed July 2020. https://www.chaipat.or.th/eng/concepts-theories/sufficiency-economy-new-theory.html.

Thongplew, N., G. Spaargaren, and K.V. Koppen. 2017. "Companies in Search of the Ggreen Consumer: Sustainable Consumption and Production Strategies of Companies and Intermediary Organizations in Thailand." *NJAS - Wageningen Journal of Life Sciences* (Elsevier) 12-21.

Tundele, S. 2015. "Eco-friendly Technology - Key for Sustainable Development." *International Journal of Research - Granthaalayah* 3 (Social Issues and Environmental Problems).

UNDP. 2020. *Sustainable Development Goals.* https://www.undp.org/content/undp/en/home/sustainable-development-goals.html.

Vapattanawong, P., and P. Prasartkul. 2005. " Future Thai Population." http://www.ipsr.mahidol.ac.th/IPSR/AnnualConference/ConferenceII/Article/Article02.htm.

Virk, A., M.A. Noor, S. Fiaz, S. Hussain, H. Hussain, M. and Rehman, M. Ahsan, and W. Ma. 2020. "Smart Farming: An Overview." In *Smart Village Technology: Concepts and Developments*, edited by S Patnaik, S. Sen and M.S. Mahmoud, 191-201. Springer. doi:10.1007/978-3-030-37794-6_10.

Wangjiranirun, W., J. Pongthanaisawan, S. Junlakarn, and D. Phadungsri. 2017. "Scenario Analysis of Disruptive Technology Penetration on the Energy System in Thailand." *Energy Procedia.* Elsevier. 2661-2668.

Wautelet, T. 2018. *The Concept of Circular Economy: its Origins and its Evolution.* doi:10.13140/RG.2.2.17021.87523. .

World Bank. 2019. "World Bank national accounts data."

World Economic Forum. 2019. *The Global Risks Report 2019 (14th Edition).* Geneva: World Economic Forum. Accessed 2019. http://www3.weforum.org/docs/WEF_Global_Risks_Report_2019.pdf.

Yeager, K., F. Dayo, B. Fisher, R. Fouquet, A. Gilau, H. Rogner, M. Haug, R, Hosier, A. Miller, and S. Schnitteger. 2012. "Energy and Economy." In *Global Energy Assessment: Toward a Sustainable Future* , edited by N. Lustig, 385-422. Cambridge: Cambridge University Press.

Publisher: Eliva Press SRL

Email: info@elivapress.com

All rights reserved

Eliva Press is an independent publishing house established for the publication and dissemination of academic works all over the world. Company provides high quality and professional service for all of our authors.

Our Services:
Free of charge, open-minded, eco-friendly, innovational.

-Free standard publishing services (manuscript review, step-by-step book preparation, publication, distribution, and marketing).
-No financial risk. The author is not obliged to pay any hidden fees for publication.
-Editors. Dedicated editors will assist step by step through the projects.
-Money paid to the author for every book sold. Up to 50% royalties guaranteed.
-ISBN (International Standard Book Number). We assign a unique ISBN to every Eliva Press book.
-Digital archive storage. Books will be available online for a long time. We don't need to have a stock of our titles. No unsold copies. Eliva Press uses environment friendly print on demand technology that limits the needs of publishing business. We care about environment and share these principles with our customers.
-Cover design. Cover art is designed by a professional designer.
-Worldwide distribution. We continue expanding our distribution channels to make sure that all readers have access to our books.

www.elivapress.com

www.ingramcontent.com/pod-product-compliance
Lightning Source LLC
Chambersburg PA
CBHW051250170526
45165CB00004B/1645